Business in the Digital, Online and Virtual Space

Julia A. Royston

BK
ROYSTON
Publishing

BK Royston Publishing
P. O. Box 4321
Jeffersonville, IN 47131
502-802-5385
http://www.bkroystonpublishing.com
bkroystonpublishing@gmail.com

© Copyright – 2021

Cover Design: Elite Book Covers

ISBN-13: 978-1-951941-97-0

Printed in the United States of America

Dedication

This book is dedicated to anyone who survived the pandemic, still in business, ready to start a business and/or ready to start again in business. You survived, now let's fulfill the dream that you had in mind, find the clients, provide the products and services that they will buy and make money while you sleep. Let's go!

Acknowledgements

First, I acknowledge my Lord and Savior Jesus Christ for giving me all of my gifts and especially my gift to write His words.

My husband who is always supportive, loving and encouraging me to utilize all of my gifts and talents. Thank you honey.

To my mother, Dr. Daisy Foree, who is my number one cheerleader and always tells me, "hang in there, you can do it." To my father, Dr. Jack Foree, who is never far away from me in spirit or my heart. I only have to look in the mirror each day to see him.

To Rev. Claude and Mrs. Lillie Royston who support me in everything I do. Especially, Rev. Royston for his careful eye to detail and his sensitive heart to content.

To my sister in law, Jean Ann Royston who didn't live to start her business, this book is in memory of you. I also honor you because you were our personal nurse and cared for all of us until your departure. You live on in our hearts but also are the catalyst to inspire someone else to survive to do their business and expand my own businesses to help and minister to the world.

To the rest of my family, I love you and thank you for your prayers, support and love.

I acknowledge every author, friend, co-worker, educator, administrator, coach, consultant, business owner, editor, illustrator and client who has inspired, educated and pushed me to be all that God called me to be. Thank you!

Table of Contents

Introduction

Throughout 2020, I heard from so many people and especially my clients that they were waiting for the pandemic to be over or the country/world to reopen or when it was safe to do in-person events to actually resume doing business. I often asked them, "What if this is it for a long while and the world doesn't return to normal? What will you do then? How will you do business or will you just shut down your business?" Now, first let me say that I am so sorry and my prayers are with any person who had to shut down a business during the pandemic. But what if the passion for your business is still there? My suggestion is you think about a way to still create, dominate, and conduct your business in the digital, online, and virtual space. Is that possible? A lot of business owners didn't think that their employees would actually do any work if they worked from home but they did. A lot of businesses did contactless, curbside, and even home delivery that they had probably never done before, but they did.

Most of the things that I talk about in this book and offer as a mere suggestion for your business I created and did on a routine basis for years prior to the pandemic. These things are nothing new, but I suggest that you find ways to at least incorporate or be an addition, an accompaniment, workbook and supplemental materials, or products and services to go along with your existing methods and resources of your business even after this horrible pandemic is over.

I have provided examples of my flyers, past events, current links to purchase my products and services and/or schedule an appointment. I created reflection pages with the symbol of working tools because I pray that this will be a working, planning, breathing and living book to help you move your business to the next level. As you know, I'm a creative so be on the lookout for a journal and planner very soon.

Take full advantage of what I have offered in this book, copy, paste, utilize, give me credit and reach out for help.

Finally, I pray for healing for those who are sick, creative ideas for those in business, strength, courage and wisdom for the days ahead.

Let's go to Work!
Julia Royston

Let's Go to Work!

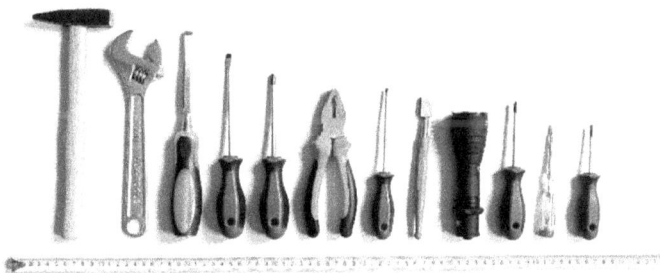

Before You Start

Doing business in the digital, online, and virtual space is not totally different from what we've always been doing. I've had eBooks and a website and provided virtual coaching and events for a long time, but now it's more prevalent than ever. I have been surprised that some people actually refuse to be in the virtual digital and online space as much as they should. I can only hope, encourage, and provide information to help with this process. Will we ever go back to the in-person events, conferences, etc.? I don't know, but will I keep moving, learning, and doing my best with the information, resources, and tools that I have? You'd better believe it. Let's go!

Something to consider before you start. The Bible says that a man who wants to build his house must sit down and count up the cost. Here we go.

The first thing is to discern what do people want or need and are willing to pay for. This is before you start creating or designing or investing or developing anything: ask yourself and ask your potential client, what do you need and what are you willing to pay for? Why? Because there may be a customer who has a need but is not willing to pay for it because they can do it on their own or don't want someone else to do it. On the other hand, there are people out there who have a need and are willing to pay for it and want to get it a certain way or to be packaged in a certain way and are willing to pay even more for a repackaged product or service. What is it? For me, someone wanted me to publish a book for them. Publishing was something that I had already done for myself with three different books. It wasn't outside of my scope of expertise, but it was different because it was for someone else. It was a need, I got paid, and I performed the service. What happened next is more people wanted the same service and my business grew. Over time, I have expanded my services and products because the needs of my clients and potential customers have grown and increased. Bottom line, I do what people want, need, and are willing to pay for it. So what do people want and need that you can and want to do?

Inventory — The new season for me started in April 2020. I had gotten out of the hospital and I was in the house. Doctor's orders were to "go nowhere" for at least six weeks. You should only go to the follow-up doctor's appointments and that is it. Stay in, rest, take meds and recover. To me, that was a death sentence, but I knew that I would not be getting much rest because I had a business to save or move and, hopefully, survive this global pandemic. The first thing I did was connect with my audience, authors of the books that I had already received deposits for and were in the works. Those clients were committed and moving forward. Thank you, Jesus! But I did take a hit and it sought to dampen me and get me down when seven clients who were about ready to start their books, cancelled. "Hold up, Julia. I'm not ready to put money into a book just yet until after this is over." Wow, what a hit to my business, but I knew that I had some books, clients, and funds that were coming in already, so I was good, but those seven clients were, in the words of my entrepreneur father, "fresh money." I didn't know where the "fresh money" was going to come from, but I kept going live on social media in my Sunday morning 10:00 a.m. slot, although the contract with the radio station was cancelled. I still promoted in my newsletter on a weekly basis, kept writing my books that I wanted to launch, launched the next writing class for 2nd Quarter, and even applied for a 501(c)(3) non-profit. I even emphasized in my weekly newsletter that since we are "now in the house, let's get those books written." It worked! Little did I know that those seven clients whom I had lost were suddenly being replaced by more than seventeen new clients and books. Yippee! Hallelujah! Praise the Lord! There were people who had waited in previous years to publish but were now ready to publish. The clients who I already had were now referring me more than ever for their friends and acquaintances to publish. I'm good. We are going to be fine.

So, **Lesson #1** — keep pushing, no matter how it looks or how scared you hard. Keep pushing. I had to keep pushing the business, services, and products that I had been pushing for the past twelve years. The message was delivered with a slight twist, but it was still the same.

Lesson #2 — I had to maintain and even intensify my drive, desire, and passion for my business. Because I had retired from teaching, thankfully, that was and could be my focus, and that was on my business. Sure, I was still taking my meds and taking more naps, but my business was the focus.

Lesson #3 — Commitment — I was committed to the business. I had worked very hard to build the business, work it on the weekends, work it in the evenings and any other free time, so I was not going down without a fight. Fortunately, my husband was in the fight with me and continued to encourage me to keep moving, keep going and said no matter what happens, "I've got your back." With that support, I'm good, but I had to have the drive, determination, desire, and commitment to the business now more than ever before.

Lesson #4 — Your mindset, focus and inner will, mind and emotions have to be fed by the right thing. You cannot have the negative naysayers, doom, gloom and "you are going to fail" in your ear during a time like this. First, when I was in the hospital, the phone connections with my husband and family were key. I could have no visitors, not even the one they allowed later in the year back in March 2020. No visitors, so the phone was my lifeline. Second, my laptop kept the conversation going with the outside world in spite of where I was. Finally, since I did have my laptop, I had all of the inspirational music, empowering word, and encouraging graphics and videos to keep me uplifted. Sure, I had my down time, but I had to tell myself, "Nope, we can't get down into any depression or feeling down right through here.

You've got to get out of here, get well, and go home to recover." Self-talk was so vital and important for me. Self-talk is vital and important to you as well. At times, it is not what the people around you say to you that stops, delays, or hinders you but it is the things that you say to YOURSELF that will stop your progress.

Lesson #5 — Look at your resources — What have you been doing that has worked in the past? Keep doing it. Still speak to that audience. Still let people know about your experience and what you have done in the past. Keep being an encouragement, empowerment, and inspiration to your audience and others. None of the Fortune 500 companies stopped promoting their products and services just added, "we are in this together" or "wear your mask, wash your hands" and supported the health crisis and cause by promoting the safety strategies and kept saying, "we are here for you" and "now you can pick up or we'll deliver or we'll ship for free" or any other new method of delivery that they could come up with."

Who do you know? Reach out to the people you already know or who have reached out to you in the past.

What do you need or what do you have that can help someone else? In preparation for 2020, I had purchased a lot of books for events that I no longer would be attending. I either would sell them or give them away. I did both. I sold some online and I gave more than 600 children's books away via my non-profit. My brand new non-profit baby actually was able to partner with some wonderful and much more experienced non-profits, which I learned from and hopefully will be mentored by in the future. I took what I had and was able to bless someone else.

Lesson #6 — Show Up — One of the things that I had to do in 2020 was encourage myself often but also encourage someone else. I posted the COVID resources available and shared the webinars to my other business owners because some were in a dark place of depression and discouragement. Sometimes the mere fact and ability to show up was able to help my business to be sustained and grow. As hard as it still is and may be more in the future, you have to go past your feelings, move forward in your faith, and show up for your business. It won't be perfect. You won't feel like it all of the time but you've got to Show Up for your own business more than others will support your business. Why? Because it is YOUR Business! You don't want to share the profits, but you want people to support you, call you, encourage you, share your posts, etc. Sometimes you have to show up for yourself if no one else shows up. I participated in events and was allowed extra time because other speakers, presenters, or authors just didn't show up. That boggled my mind, but I just enjoyed the opportunity and took advantage of it. I didn't bribe anyone, step on someone else's toes, or do anything underhanded to get the opportunity; I just showed up. Show up for your business and see what happens, even after the pandemic is over, show up and show out!

Lesson #7 — Follow-Up — When I first began my business, I had one publishing client with one book. Hurray! I had published three books of my own within the first two years of my business. When I began to say "yes" to more clients, the need for follow-up began. With fewer than ten clients, it was fine putting them on a calendar and texting, emailing, or calling them directly. But as my business has grown, it's time to either add help or develop a system for follow-up. It is critical. I admit I have lost some people along the way because my system of follow-up hasn't been perfect. I have control issues, which I admit, but over time, I have to add a system and people with that system to follow-up. You do, too. Some people are not ready to sign, schedule, or do business with you right away, but if you don't stay in their face via a newsletter, email, and/or social media, they'll find somebody else to do business with. There are those, which I'm thankful, who, when they are ready to do business, want to do business with me. Thank you again! But there are those who find other businesses, people, and opportunities, which is what I am striving NOT to have happen. It can't be helped all of the time, but I want to minimize the ability to work with someone else as much as possible. So what do you do to follow-up with people whom you've met as well as those whom you have done business with in the past? Postcard, email, text message, social media, private Facebook, membership, mentorship, and just a phone call are just a few of the ways to follow-up. The system has to work for you, but it also must work for the people you are attempting to attract and do business with. It can't be just the way you want to follow-up and communicate but also the way they respond and hit that sign me up button. It will work if you work it. Follow-up, follow-up and follow-up some more.

Reflection

_____

_____

_____

_____

_____

_____

_____

_____

_____

_____

_____

_____

_____

_____

_____

_____

_____

_____

_____

_____

_____

_____

_____

Systems

Communication

Prior to even starting my publishing company, I produced a monthly newsletter. I performed at multiple churches, events, and conferences so when people purchased my CDs, I asked for their email addresses and began to send them my calendar, any new music that was upcoming, and/or any other things that I was doing with my music. I did that for approximately ten years or more. So when I started my publishing company in 2008, it was a no brainer for me to communicate with my current email list about my new company, new products, and my new calendar. Communication has been the key for me to stay in connection with my audience even if they are not interested in what I'm selling; they can share the information with others. It has worked extremely well for me.

When March 2020 was in full swing, I participated in as many free webinars to assist businesses as I could. I share the links with others and the first thing that every speaker said was to, "stay in communication with your clients, customers, and/or potential customers." Don't abandon them. Don't stop checking in with them, and, if appropriate, promote a very needed product or service. I hesitated at first about promoting something to sell but realized later that although people were working from home, they were still getting a regular paycheck. Some people had just moved locations but had not lost their jobs.

I went live on social media, I posted on social media, I continued to send out my newsletter, and sometimes I blogged. I realize that even during this time, communication was one of the major keys. Communicate your sympathy and empathy for the time that we are in together. Communicate a service that you might have not offered before such as delivery, pick-up curbside and/or contactless, or a new product that might have helped or assisted people to live better lives during the crisis. No matter when and if the global pandemic ends or never returns (we are hopeful), always communicate with your clients. Establish, maintain, and, if possible, increase your communication systems. What is your communication system? Need help? Reach out at www.connectwithroyston.com.

Access

Calendar Is My Life

Scheduling your day has become unprecedented during the pandemic. You wouldn't think it would be necessary because you are staying in the house, not going anywhere, just getting clothed from the waist up and then go, right? Wrong? Being able to allow opportunity for you to have time for your clients, time to plan time, to think, time to rest, time for a break, and then time to get back on that horse and do more work again. So I had to determine my day had to be planned exactly like it was planned. When I actually worked a nine to five job, I could not run 24 hours a day. And I had to schedule a time to eat time for a lunch break, time for a snack in the morning or my breakfast in the morning, time to fold clothes, time to take a shower, time to whatever I needed to do and have client time. So your schedule or your calendar was essential — is essential — not only in a pandemic, but also in an online digital setting.

You've got to make sure that you have a schedule of when and where and how. And I do have a calendar of a word will, can schedule time to talk with me, but then you also have to make sure that you're consistent and going back and blocking out time. When either you don't have a client call or somebody in boxes, you gotta go and say, "I can't do it at two o'clock. Cause I told them in the inbox on Facebook, I would talk to them at two o'clock because you have the potential." When you put out the calendar for everybody, you can be overbooked and double-booked. I found that out the hard way. And then there are other things, such as webinars that provided business information I needed to find out. There were times I needed to be trained, tooled, and reskilled to find out more important information.

So the schedule is your life and key to being organized even in the digital, online, and virtual space. You've got to go by a calendar. You've got to go by a schedule, and then you can't schedule your time with meetings so close together that you're not able to, um, make a summary of what was talked about. You can't follow up and have what's next steps. I'm trying now to, after I have a conversation, make notes as I have the conversation, but also have time to follow up before another call comes. And if somebody is calling with the "Hey girl, what's going on?" they don't get responded to. "I'm sorry, love. I'll get back to you later on at another time. What you doing? I'm working." They know I'm working; everybody knows I'm working. So therefore, you have to have time for family, and time for friends, and time for those, "Hey girl," catching up conversations and "How you doing," so that's during the break time, that's during the lunchtime, or that's during the afternoon time before you take a nap, but if your body starts screaming, I'm tired. I'm exhausted. Lay it down before you get sick and something else happens.

Schedule Calls

Tips that I had to learn in the digital, online, and virtual space showed I have to space them out, take breaks, and then maybe even schedule more on another day. I can't talk five hours each and every day, and that work for me. It's too much. So my suggestion is to have breaks on your calendar where no one can schedule a call. For example, one hour break and then another hour to schedule, and then don't forget lunch breaks and even a nap some days. Spain knew best with the afternoon siestas.

Virtual Assistant

Whether the assistant is virtual or in-person, you will eventually in business need some help. I suggest that you know as much about your business and all aspects of your business in the beginning. But over time, some aspects are repetitious daily or weekly administrative tasks that are not confidential but proprietary that could be handled by someone else. I have control issues as a business owner, but if I am going to live, expand, and increase my business, I am going to need help. So I had to consider a couple of things before hiring someone.

1. What exactly do I need done?
2. How much budget do I have to spend for these tasks?
3. I need to hire someone on a project basis and NOT on a long-term, employee basis. I will submit 1099s to U.S contractors but don't want to pay benefits, taxes, etc. So I have to talk to them about being a contractor rather than an employee. So I have to ask them when they are available, what their pay rate is, and how many hours a week they want to be paid. I see if

those pay rates and hours meet my needs and if not, move on to someone who will.

4. Finally, you have to make sure that not only are you financially ready to receive help but mentally and emotionally. I had to check myself regularly about this because I CAN do these things, but now I DO NOT have the time or schedule to be able to do these tasks in a timely manner. I prefer missing some essential tasks to my business than affecting whether I am more profitable and productive in business or have to go out of business. It is a choice. With my ego in check, I move forward with a person whom I trust but have to make sure that those tasks get done and I pay them.

Payments

Five to ten years ago, how to get paid wouldn't have been a big issue. It was either cash or check, and you either got the money in person or very few smaller businesses were accepting money online, over the phone and/or from a website. Today, in 2021, you are less likely to accept payments in person and you are more likely to be accepting electronic payments. The times have changed and probably will take a while before they return or if they ever return.

Additionally, another reason some businesses have not grown is that they have limited ways to accept payments for their products and services. If there is another business who provides the same product and/or service, you may lose a customer based on how you deliver the product and how you accept payments for the product. It is the difference of money or no money and/or sale or no sale. Throughout this book, you need to realize that times have changed and as a business owner you have to prepare to change with those times.

The various ways to accept payments are as vast as the number of companies who have gotten into the money acceptance, money lending and borrowing business. It is amazing. Below are just a few of the payment options that I have used personally:

- Take the credit card information over the phone
- PayPal
- Squareup
- Venmo
- Zelle
- CashApp
- Stripe

These are just a few of the payment options but this doesn't include all of the merchant accounts and plans that you can offer via your bank and/or credit union. The options are limitless, but here are a few tips for having these multiple payment options.

1. Good relationships with banking institutions
2. An active and in good standing bank account
3. Sales more than fees

These are just three of the main issues to keep in mind when conducting business. My father used to say that some people are not in a position to be in business. It takes not only a business mindset but foundation tools and systems to be in place to be able to stay in business. In times past, it didn't matter how good your credit rating was or if you had a bank account because we were in the cash and carry days, but now, your credit does matter and the ability to accept electronic payments will determine whether you stay in business or go out of business.

Make sure that you evaluate your personal spending habits, your current income versus your expenses to determine if there are some things that must change before you even start in business. I confess when I was younger I had bad spending habits. In my thirties, God began to deal with me on a serious note of coming to a right relationship with money. I am not writing about something that I know nothing about. I had to get my financial mindset and habits to align prior to starting my own business. I had to stop shopping for what I wanted and then realizing that I might not have enough for my needs. I had to move away from people and say "no" to some other things to stay on track for my business, career and life goals. These are things that you really have to consider prior to going into business and then not falling into the statistic of being out of business within the first three years.

Additional Systems

A few other systems that we will discuss later are the online store, websites, and events. Be ready to take notes. Let's go.

Digital

Space

Digital

"involving or relating to the use of computer technology"

www.dictionary.com

Product Creation

One of the biggest questions that most business owners ask is, "What will I sell?" In the digital, online, and virtual space, what will you offer? Due to the pandemic shut downs, limits on how many can gather, and the social distancing demands, having physical products that you can ship has been challenging, but what about digital products? Digital products are those products that can be sold electronically via email or download. There is no contact with these products. They can be distributed automatically and you can make money while you sleep. The music industry has understood how powerful and profitable downloads and streaming of music has been critical to hopefully, the sustainability and the delivery of music to a broader audience. In recent years, the ability to fill stadiums, music halls, and other large venues so that the physical CD and other merchandise can be bought with the face and logo of the artist has been non-existent. Digital downloads are now the #1 way for music to be distributed.

What digital product can you produce for your business? Here are just a few of the digital products that I suggest for you to produce and distribute.

eBook

Some years ago, there were a number of articles stating that the eBooks would replace the printed book. It is not so and hasn't happened at the release of this book. What has happened is the need for an eBook to not only accompany your print book but to introduce your business and/or other products and services. I do believe that you will always need a book, but books need to be in multiple formats to reach multiple audiences. The ability to have a book on your computer, laptop, tablet, and phone, as well as in your hand, to read on the beach, is vastly growing. Presently, in my business, all of my book publishing packages include both a paperback and electronic format. It is now standard and not an option. I believe that the more media formats that your message is available in, there will be more opportunities for a larger audience.

eBooks can be sold on your website, along with a class, workshop or conference or simply to produce something short enough to be a lead magnet, which is a high quality content eBook that will be given away. If you are new or establishing your credibility in your industry area, people are more likely to give you their email address in exchange for your eBook rather than purchasing the eBook. Over time and your experience, the eBook can be definitely sold or bundled with another product or service. In addition to just receiving email addresses so you can communicate with these people in the future, they get your knowledge and insight into an industry that they already have an interest in.

For more information on creating eBooks for your business, visit http://www.bkroystonstore.com and sign up for the "eBooks for Business Course" that is available for independent study and on demand.

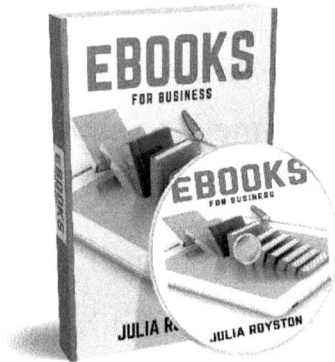

Workbook

The first word in "workbook" perfectly describes for me what a workbook is, a working book or something that you are working through to achieve or complete a particular goal or project. I have written several workbooks, including a workbook to teach and walk you through exactly how to write a book. I have written a workbook that accompanied a devotional and main book so that the instructor and/or reader can dive deeper into the subject matter, reflect, create, decide, and come out with a great understanding of the subject after having read the text. You can sketch, brainstorm, and work out a lot of problems and issues before you start a project in the workbook.

If you are a financial investor or represent real estate company, workbooks are perfect for working through and getting a better understanding of the numbers.

For more information on how to create your own workbook, reach out to us at http://www.talkwithroyston.com. To purchase any of Julia Royston's workbooks, they are located at: http://www.juliaroystonstore.com

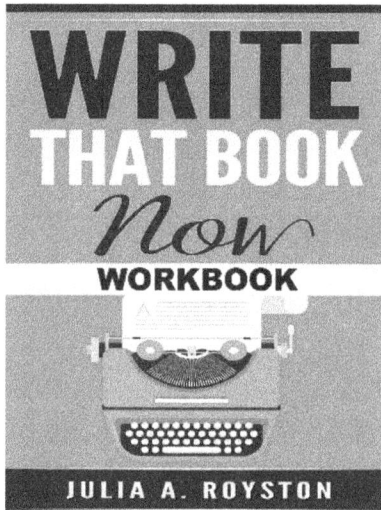

Checklist

In certain businesses and industries, there are pieces of information, documentation and reports that are needed throughout the process. If the client had the documents or information handy, don't you think that the process would move much smoother? Of course it would. Additionally, there are people who know that they need a will, power of attorney, or a house, but they don't know what information is needed on their part to begin or complete the process. The fear of the unknown or the rejection that they don't have the right documents or money needed or whatever is necessary at times, keeps people from getting the help that they need from you and the service/business you provide.

Ease their minds and bring clients to you faster and easier by providing a checklist with the information that they need to get started or complete the process. Get their contact information, so that you can keep in touch with them while they are gathering the information. Your help with the free checklist will add value to your business with the information you provide but will also make you and your business a valuable and "go to" contact when they are ready to go to the next level.

For more information on creating checklists for your business, schedule a time at http://www.talkwithroyston.com.

Checklist for Publishing

Front Pages or Front Matter

_____ Title Page: Title of the Book Subtitle (optional) Author's Name Publisher's Logo Illustrator's Name goes

_____ Copyright Page – Publisher Information – Copyright Year – Copyright Statement All Rights Reserved, Cover Image/Layout DBN, Printed in the United States of America – Illustrations by, Scriptures copyright content goes, Other Permission.

_____ Publication Reference List or Other Books by the author or Pre-requisites or Additional Books in the Series

_____ More by this author

_____ Special Dedications – that someone was deceased

_____ Dedication Page

_____ Acknowledgements

_____ Table of Contents

_____ Foreword – Written by someone else that is a book endorsement

_____ Preface – similar to a Introduction but not a requirement

_____ Introduction

_____ List of characters and/or Family Tree Related to Manuscript

_____ Second Title Page

Optional pages or documents that have been added but not the norm

Alternative Front Pages – Very optional and very few times ever seen used.

Resume

Links to Videos/Other Advertisement Materials/

Planner

Planners are similar to workbooks in that they are a way for your potential client or current client to put their ideas, plans, numbers, and future on paper. Writing it down on paper or on your computer and seeing it with your own eyes and sometimes saying the dream, goal, or objective out loud is a way to instill the idea not only into your brain but also into your daily work.

The person who drafted, designed, and delivered the planner will be forever associated with the idea or planner as well. Thus, when the next steps, next project, or a referral for your type of services is desired, you should make the list. Helping people to fulfill their dreams is so instrumental to your business, its potential, and next level growth. Why? Helping others will help you. Not only to produce profits in your business if people do business with you, but also to increase your relationships if they are satisfied with your services. I am a witness that people will refer you to others even if they haven't worked with you themselves. Help someone to plan out their life to succeed and watch your life and business succeed as well.

If you need help with a planner, let's have a conversation at www.talkwithroyston.com or purchase any of Julia's planners at www.juliaroystonstore.com.

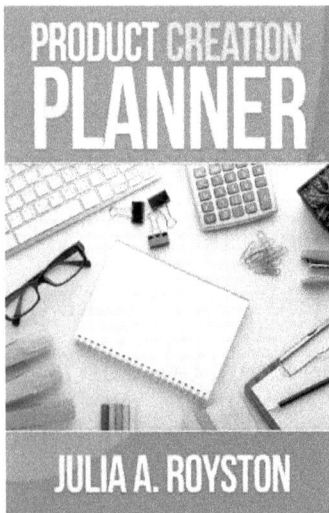

Journal

I wrote my first book from the simple spiral eighty page notebooks from the discount store. I just wrote what was on my mind and in my heart. So, the practice of journaling is very therapeutic and helpful for the mind, body, and spirit.

You, as a business owner/author, can create your own journal that can contain blank pages with a decorative cover. Other types of journals can accompany devotionals or inspirational writings where people can write down their thoughts, reflections, and directions for the next steps based on the devotional lesson.

You can also create journals that accompany a process, strategy, project, or method of creating something. For example, you can create a weight loss or eating or food journal for a membership program. If you are a realtor, you can have a new home owner journal for the process of owning and moving into their first home.

There are so many ways to create, use, and bundle journals with existing products and services. Create your own journal today.

The latest inspirational journal from Julia Royston accompanies her Queen inspirational book. Purchase this and any other Julia Royston Journals at http://www.juliaroystonstore.com.

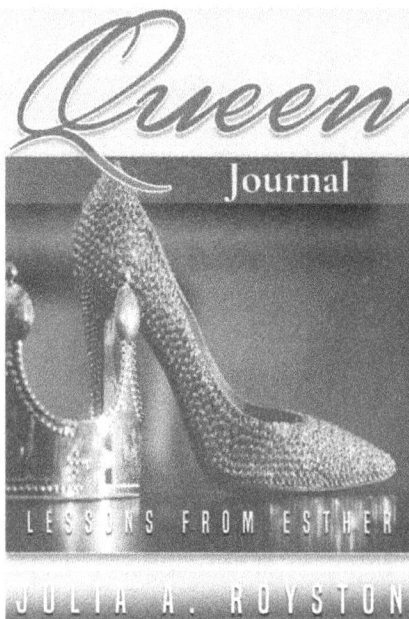

Informational Graphic

Now first, an infographic or informational graphic is a visual, informative graphic that provides specific points and information regarding a book, program, topic, industry issue, or other project. It is usually very colorful, distinctive, clear, and to the point. Second, there are some major points regarding your products or some important topics or things to remember when working with your business that you should include. If you can make the graphic with the colors of your brand and associate it with a product that your business produces, that would be even better for people to remember.

Here is an example of an information graphic:

QUEEN BEHAVIOR

HTTP://BIT.LY/ROYSTONQUEEN

PEOPLE
A Queen rules but the people she surrounds herself with will make all of the difference in her reign.

PROBLEM
A Queen finds, creates and delivers solutions when problems arise instead of creating more problems.

PLOT
Plots to sabotage, discredit or tear down are not tolerated by a Queen

PLAN
A Queen plans her work and works her plan to accomplish great things.

PERFORMANCE
Plans are wonderful but it's the performance of a Queen that truly matters and will make a difference

Videos

Videos clearly fulfill the saying, "a picture is worth a thousand words" to the 10th degree. Today, that same picture now has animation, sound, other images, text, and music added to make the picture even more appealing and interactive. A video grabs the viewer's attention. A link added to that same video at the beginning, middle, and end can add money to the pockets of the producer of the video. Videos should be authentic and don't have to be polished to be effective. We have seen how horrible videos showing murders of innocent people have sparked protests and riots in the streets across this country. We have seen dancers, singers, and other artists display their talents and land multi-million-dollar contracts and become world famous.

Educators, celebrities, coaches, ministers, organizational leaders, business leaders, and trainers provide video courses and become famous, profitable, and transform lives around the world from their videos. Memories are made, savored, and cherished through video. The video commentary surrounding the message of your book should be the introduction to your website, course, or program. There are so many ways that video can be used with your book, business, and product line that it is a crucial and critical piece of the business equation. It is no longer an option, but a necessity. Whether you utilize several hundred-dollar HD cameras or your phone, make it meaningful, content-rich, and come straight from the heart and watch how it will reach the masses.

To subscribe to the BK Royston YouTube Channel to see Julia Royston's video collection, visit http://bit.ly/bkroystononyoutube.

YOUTUBE CHANNELS

Julia A. Royston
BK Royston
Publishing

Follow Both Channels and Subscribe!

SUBSCRIBE

Audios

Audio files are extremely popular, especially for busy people and those who enjoy listening to and learning new content while working, exercising, and driving. Because of the recent pandemic and quarantining, audio content demand and need have increased. People can be inspired, educated, and entertained by the audio files that they listen to.

You can turn these same audio files into an audiobook (which we will talk about more). High quality audio is necessary for exposing your business through the outlet of podcasts.

How do you create an audio file? I use an app on my phone to create audio files. It is economical and the app provides a high quality sound. It is all quick, easy, and in an extremely quiet space, can be recorded and uploaded immediately to social media, my website, or sent directly to customers, members of a membership or mentorship group, as well as added to a blog post or other paid program.

If you don't know what to say, write it down. Practice it multiple times. Speak slowly and clearly into a high definition device in a quiet space. Also, repurpose previous content that you created in an eBook or blog and bundle it for purchase in your own online store or a part of merchandising for your course or membership group. In a few steps and a few minutes, you have an audio file that can motivate, stimulate, and inspire millions.

If you have questions or need help, reach out and let's have a conversation at www.talkwithroyston.com, or listen to some of my audio files at www.juliaroystonstore.com.

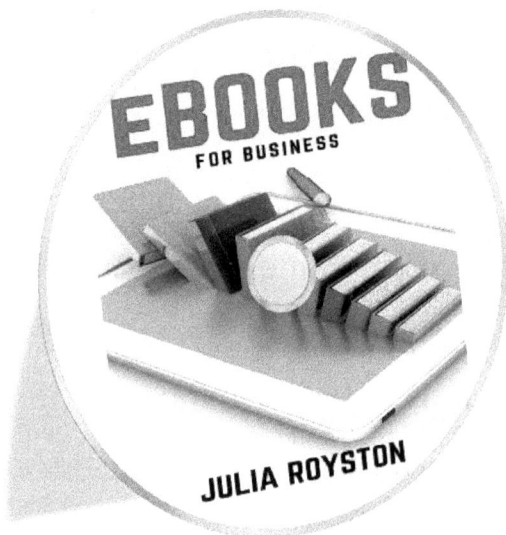

Audiobooks

An audio book is the audio version of your entire print book or eBook. You can develop an audiobook with just three to five chapters, but you need the entire book from the title page to the end matter of the book.

For audiobooks, it is an investment in this media format. I advise you to make an investment in a recording studio, with an experienced engineer, quality recording, editing, and production. If you are going to complete multiple audiobooks, you may invest in the equipment to be able to record in your own home. My husband does production for me with regards to editing, etc.. I have the equipment to record in my own home.

If you would like more information or the list of the equipment necessary for audiobooks or podcasts, visit https://gum.co/NEcqoW.

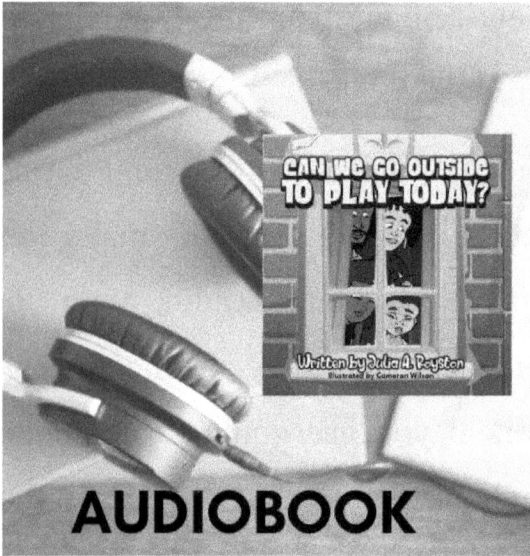

What Digital Product Will You Create?

_____

_____

_____

_____

_____

_____

_____

_____

_____

_____

_____

_____

_____

_____

_____

_____

_____

_____

_____

_____

_____

_____

Reflection

_____

_____

_____

_____

_____

_____

_____

_____

_____

_____

_____

_____

_____

_____

_____

_____

_____

_____

_____

_____

_____

Online Space

Online

"controlled by or connected to another computer or to a network"

www.dictionary.com

Show Up Online

I hear so many people say that they are not active online or spend time online, but they still want to know what's going on online. With the limited in-person capabilities, online is no longer an option but a necessity, especially for those who have a brick and mortar business, a non-profit, or ministry.

If you are going to do business with, minister to, or connect with people in this day and time, online is where you have to be. Now, you don't have to be like me and be everywhere, but you need to show up somewhere, consistently, and socially sometime. There is so much to learn in the online world and so much to share, empower, and support in the online space.

Here are a few suggestions of places that you need to show up strong in the online space.

Social Media Presence

When people come to me and ask me about publishing their books, helping them with their business or organization, I usually go to social media first. I want to see what they are telling, showing, or displaying for the world to see. Social media is where I conduct business. I display some personal aspects of my life, but my primary goal on social media is for people to get to know me, my business, my products, and my services.

Now I know you are asking yourself, "Why social media? It's not a product or service that I own." True, we don't own social media but we can present, promote, and profit from being active on social media with our products, services, business, and organizational resources.

With that being said, pick two. I am not talking about the lottery numbers. I am talking about two social media outlets. Two outlets that you will be diligent in learning about, posting, and engaging with others to prosper and move forward with your book, business, and organization. That's it. It is no longer an option but a requirement.

I don't know if you realize it or not, but some companies view your social media page before they hire, promote, or interview you. Be careful what you post, but also, I have had event promoters not allow me to participate in an event because I didn't have high enough social media likes and followers on a certain outlet. It does matter.

My last soapbox about social media is if you are participating in an event and they have a flyer or webinar or video, post it, share it, and spread the word about it. It helps you and everyone else. Just a hint, people will want to do more business, collaborate, and partner with you when you share.

Here are my social media outlets and handles. Follow, Like and Share.

Facebook: @juliaaroyston

Instagram: @juliaaroyston

Twitter: @juliaakroyston

LinkedIn: @juliaaroyston

TikTok: @juliaroyston

Media/Broadcast

What started out as a podcast was picked up by an Internet radio station and transformed into a weekly broadcast. My broadcast is "Live Your Best Life," and it is a show where I present guests who provide information, entertainment, and empowerment for people to actually live their best lives. I wanted to keep the topics open and not just focus on writing or publishing but be able to talk about anything that interests me and my listening audience. Fortunately, when I travelled, I met so many people who had stories and information to share, not to mention my publishing and coaching clients and friends, who are so willing to be my guests. I am very blessed. I was approached with the idea and gladly accepted because it helps others to be their best self. It also introduces more people to me, my business, and my clients. I monetize my broadcast with commercials and interject my music. I am living my best life, so I want to help others do the same thing.

For more information or help with a broadcast, schedule an appointment at www.talkwithroyston.com.

Podcasting

Podcasting has become one of the fastest growing ways to communicate to an audience. We have talked about audio and video earlier, but with my podcast, I have combined both. I stream my broadcast, "Live Your Best Life," through an Internet radio station, www.envision-radio.com. I also have a podcast, "Book Business Boss Podcast," that is heard where podcasts are distributed. I also video tape the broadcast and podcast, so that I can use it on my website and/or YouTube channel. I told you earlier that I strive to repurpose content as much as possible. Podcasting also is a way to have a broadcast/telecast that can be repurposed on other Internet-based platforms for syndication, as well as used again on traditional radio as pre-recorded content. Content is king. Adding noteworthy and informative guests can only enhance your podcast. For me, it is a way to train and teach authors, business owners, and other organizational leaders how to interact with the media as well as give them an outlet to promote their upcoming releases, events, and other products and services.

For more information, let's have a conversation at www.talkwithroyston.com.

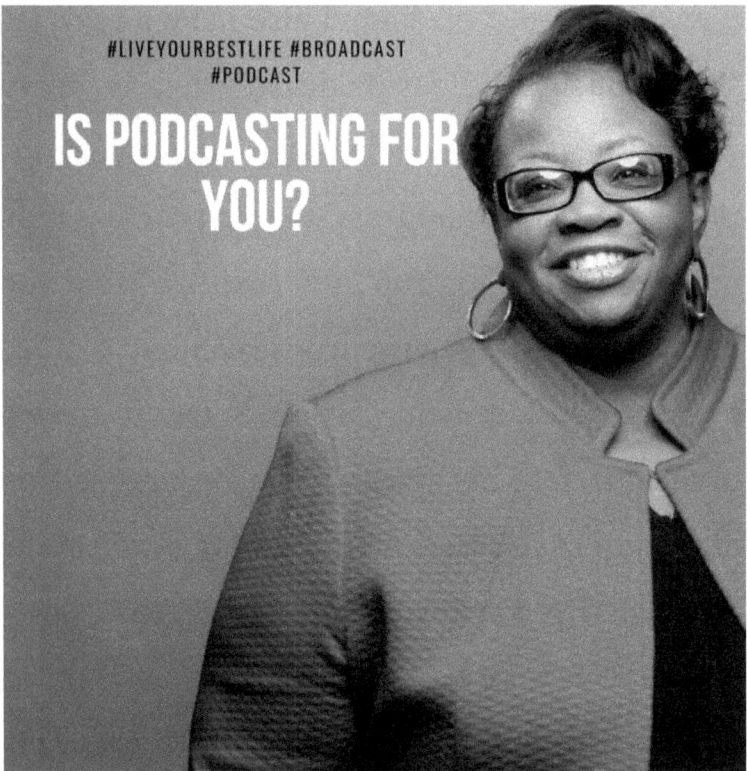

Book Business Boss Podcast

In the Spring of 2021, I established the Book Business Boss Podcast as a 15-minute podcast to talk to authors and their books. It is a 15-minute podcast that I will use on my podcast and YouTube channel for marketing and promotion of books. Additionally, the author can share the link for examples of their previous interviews, embed them on their websites, as well as use it as a training tool for better future interviews.

If you need help with developing a podcast, reach out at www.talkwithroyston.com.

Main Stream Media

What does mainstream media look like these days in the digital, online, and virtual space? Television is still king, and the ultimate goal for anyone who wants to speak to a global audience. During the pandemic, international television broadcasters had to have their anchors broadcasting from home. We saw not only green screen backgrounds but actual books on shelves or desks or kitchen counters displayed while they were giving the news or commentating on an event. I thought to myself, if they are using web-based interfaces to communicate, so can I. I might have been going live to my social media audience, but I was still in the Internet space. I might not have been a guest on one of the national stations, yet, lol, but I was out there sharing my content, products, and services.

What's next? I noticed during the pandemic more and more streaming, Internet radio, and Internet broadcasting outlets rising. I also noticed that the mainstream media outlets had to go to WebEx, Zoom, and other streaming outlets to be able to broadcast with multiple people safely. So the opportunities for more independent media companies have definitely increased. The biggest issue that I've seen with these streaming and other media outlets is content. So what do you know? What are you an expert in? What can you do and teach others how to do? Get your set, script, guests, wardrobe, and lighting together because there may be a media outlet waiting for you. Are you ready? Be ready because someone could be jumping in your inbox, sending you an email, or sliding in your DM very soon.

This is the main reason why I went from going live every day on Periscope to a podcast to a weekly broadcast. I'm not the only one who can have content and guests; you can, too. Need help? Let's talk at www.talkwithroyston.com.

Websites

Why would I need to mention websites in this book? Because people still do not have a website in business or maintain their websites. A website is like anything else in business, it has to be maintained, be cared for, be cleaned up and have content in constant creation to remain relevant. I wish that you could design a website once and never return to it again, but that's not so. A website grows just like your business grows. A few tips of why you need a website:

- Internet Real Estate
- The Vetting Place
- You Control the Content
- Headquarters

I know that I have said that you need a website, but what should actually be included in a website? I'm so glad that you asked.

- History of the business — About Us
- Location, if there is an in-person location
- Who is on the Team?
- Testimonials
- Products
- Services
- Past Events/Current Events
- How people can contact or communicate with you
- Graphics/Logo/Images/Pictures
- Introductory Video

Non-Profit Website

Next, I applied for a non-profit organization and that needed its own branding. So I designed a website for that to keep it totally separate from my other businesses. It is a business that should and does stand on its own.

www.bkroystonfoundation.org

Business Websites

Finally, when the pandemic hit, the first thing I realized was that I needed more websites because I was going to expand my business. I had eCommerce stores on another platform, and they changed ownership, and so did I. I wanted my own ownership of my content, inventory, and sales. I created stand-alone online stores so I could combine my inventory, link, and point the store page to one place, and even have a resource for my book fair store and any other supplemental information for all of my brands.

www.bkroystonpublishing.com

www.juliaroyston.net

www.juliaroyston.net

Celebrating
5 Years of
Great Stories!

www.royalmediaandpublishing.com

www.royalmediaandpublishing.com

Online/eCommerce Store

Who wouldn't want to have their own store that they can sell merchandise anytime day or night? You don't even have to be awake to sell your merchandise. When you wake up, someone has bought something, you have a notice of the sale and when the shipper will deliver it, and you enjoy the profits from the sale. Easy, right? Not so fast. Just like any other retail outlet, you have to have inventory that people want to buy, and you have to have a promotional strategy to get in front of potential customers. The biggest consumer of your time is getting/creating the merchandise and organizing your store. There are some "made for you" eCommerce websites that can help with the organizational part, but you have to create the products. After that, the next biggest part is promoting your store. As you can tell, I have been promoting my stores throughout this book. That is only one way to promote your store because as many people as I would like to buy this book, all may not want the information in this book, so they won't see my ads for my store. I promote my store on my business cards, blog posts, newsletter, magazine, social media, and my websites.

Keeping more of your profits is a primary reason I have my own eCommerce store. Why? When you have distribution through someone else, no matter how large the distribution outlet, they require and take a portion of your profits as a distribution fee. Rightly so, because they have bills, expenses for upkeep, shipping, handling, and taxes that they have to pay as well as the distributor. On the other hand, if you have your own store and have created digital or physical products, you don't pay the middle man except what you pay to have the store access. Most of the time, there is no fee that you would pay unless someone buys something from you and that is primarily because of the credit card fees. Credit card fees are still less than distribution fees, so you, as the owner of the store, are able to retain more of your profits when you have a store and sell directly to your customer than if you went through a distributor. You still have to promote to your store and the distributor, but if you promote to both, you double the exposure and double the profits. If you need help selecting or putting up your store, reach out. Let's go!

ROYSTON ROYAL
BOOKSTORE

www.roystonroyalbookstore.com

www.juliaroystonstore.com

Courses

A workshop is defined as "a seminar, discussion group, or the like, that emphasizes exchange of ideas and the demonstration and application of techniques, skills, etc." (dictionary.com) Now that you know what a workshop is, imagine taking several workshops or expanding the workshops into a multi-week course. The best courses have come as a result of teaching on a topic and people wanting to know more and dig deeper into a specific topic. The beauty of being in the online space is that the platform expense or the hosting software of the course and your time is the majority of where the expense comes from. The products should be in digital format and distributed digitally as well. The biggest issue is finding people who want your course and are willing to pay for it. In the beginning, we talked about creating and designing products and services that people actually wanted and were willing to pay for. You may have two, three, or four, but what if you were able to find five more? That would be great!

First, if a topic is something I know that people are interested in and that I have a knowledge of, I usually write a book about it. The books I write are directly related to my business, purpose, or interest, period.

If I've written the book, I usually teach from my own books. Why? Because I know the content because I am the author. Second, my book is now a textbook that goes directly with my course. Finally, I receive the profit from the book and the 6, 8, 10 or 12-week course.

It has been said that people will pay you for what you know. I also say that you have to be ready to charge them and have the content formatted so that people can be plugged directly into the course quickly and easily.

What do you want to teach people? If there is a specific product that we want to create, a multi-week course is probably the best way to do it. I have a 10-week course on how to write a book. I have a 5-week course on blogging. So again, what do you want to teach people? Always have an end product or end result in mind when you are teaching the course. Have a lesson plan, activity, and homework for each week.

How long will it take before they receive the replay? Will the replay come through email or another large file transferring service? How long will it take for feedback to arrive about their homework assignments? These logistical avenues need to be worked out prior to beginning the course because the questions will be asked. Additionally, if necessary, a virtual assistant may have to be retained so that people receive responses in a timely manner if your schedule does not allow.

For more information on how to create a course, visit http://www.bkroystonstore.com and you will find a course already created on "How to Turn Your Book into a Course." Of course, there is, why wouldn't there be? Lol.

Boot Camps

When I think of boot camps, I think of the military. The boot camp is filled with intensive, high quality, and packed information with materials, handouts, and examples to get the most out of the experience. I normally do boot camps for four weeks. The training sessions normally don't last but one hour. The key to the success in a boot camp on the one hand is that the instructor is organized and distributes information in a timely manner, but on the other hand, the participant must be ready to do the work. A 4-week boot camp is different from a 10-week course because you have time to catch up, but if they don't stay the course, the four weeks will get away from you and sometimes the participant will blame the instructor for not obtaining the results that they felt they were supposed to receive. I advise you that if you are planning on doing boot camps, emphasize in the logistics or pre-bootcamp work that their participation is key to success.

Online courses are wonderful for both parties, but, just like a gym membership, people have to incorporate the activity into their life or they will be paying for something that doesn't become a part of their lives and then doesn't give them the return on their investment.

Masterclasses

A masterclass is an online class but it is a one-time, intensive training that lasts between one and two hours. There are some higher priced master classes that last longer than two hours, but two hours is about my clients' limit. My ideal time would be 90 minutes, but it all depends on the skill of the group that I'm teaching. In the online space, you have to be careful that you offer the topic with a timeline that is conducive to your clientele at a price that they are willing to pay.

I find the sweet spot for pricing for master classes is $49 for me, but it all depends on the topic, the time, and who is interested in your class.

Because of the price point of my business, I offered the first master class concerning having your own online store for free. This gave me an opportunity to work out the bugs, get feedback from the attendees, as well as refine the content for the future classes. They enjoyed everything, but it was my first time and I wanted them to be honest.

I have since taught the same master class four more times at a fee that seems to work for everyone in the class. I sent the attendees the replay but will record the content alone for my on-demand and independent study store.

I didn't use any of the recordings from my master classes because the time was so intensive and short that I allowed the students to ask questions along the way and not wait until the end of the entire class.

So you need to ask yourself, what do my clients keep asking me about that they say they want or need and I can teach them?

The first thing to seriously remember is that all of these workshops, courses, boot camps, and on-demand courses add value to your business. Second, it gives credibility to you as an expert in a particular area of field. Even if you've never taught it before, after that one time, good or bad, you are an expert. After you record it, you can learn from your recordings and do even better next time.

Finally, if you charged for any of the classes, you made money from what you know and not just from what you can do. I've been encouraged by other coaches/business owners/entrepreneurs to make sure that you have opportunities for people to pay you for what you know and not just for what you can do, which is labor driven.

People will pay you for what you know, and if you can do it from home in a virtual space, it's even better.

For more information, let's have a conversation at www.talkwithroyston.com. An example of one of my Masterclasses is below.

Instructor: Julia A. Royston

Create Your Online Store

NEXT MASTERCLASS

JANUARY 25, 2021 7:00 TO 9:00 P.M.

www.bkroystonpublishing.com/classes
$49 each session - No Refunds

Comments from Participants..

"Amazing Training!"

"Great Nuggets Taught Here!"

"This class was excellent!"

"Answer to Prayer!"

REGISTER TODAY!

On-Demand Courses

As much as I love doing live classes, there has arisen a need for independent study or on-demand courses. Since I record every course that I teach, it is quite easy to package the courses and resell them. After two to three times of conducting a course, I feel comfortable with the class and its content. I do one of two things: take the best of the editions of the course that I've already done and package them and sell them OR I record the content again, package it, and then sell it. I confess that the majority of the time, people want to be in my live classes rather than watch the replays, but I have them already packaged and on a website for easy payment and download just in case. I do not use anyone's ideas, faces, etc. I usually have a greeting and Q&A after the training, but I only resell the trainings themselves. These are recorded without video for me or anyone enrolled in the class. The participants' identities and their questions, etc., are all proprietary and kept in the strictest confidence.

As the world continues to be more and more comfortable with the digital, online, and virtual spaces, having that on-demand, independent study learning is helpful. The on-demand courses are deeply discounted, but also included in each course is a one-hour of coaching with either me or a member of my team. This is an opportunity to ask questions and get any obstacles that they faced during their self-study course worked out. This can be scheduled at the completion of the training course and at the convenience of the student and the instructor.

For more information about on-demand courses or to schedule your class for extended learning, visit www.bkroystonstore.com.

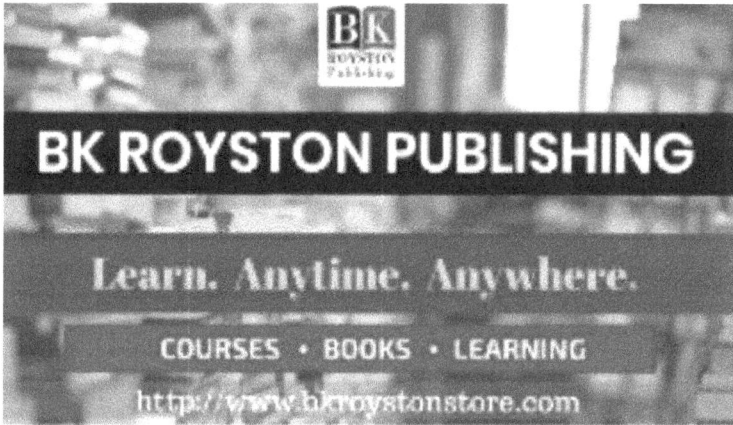

Reflection

_____

_____

_____

_____

_____

_____

_____

_____

_____

_____

_____

_____

_____

_____

_____

_____

_____

_____

_____

_____

_____

Virtual

Space

Virtual

"carried out, accessed, or stored by means of a computer, especially over a network."

www.dictionary.com

Virtual Communication

I must confess that this is the part of the digital, online, and virtual space I had to get used to and, with the pandemic, was furious and somewhat scared of how this would impact my business. I really like to travel and meet people face to face. I enjoy eating a meal or appetizers and discussing books and possibly how we can do business together. There has always been a great experience at a conference or event of more than 1,000+ people and realizing that I had a great potential of making connections to future published authors and clients of mine.

In March of 2020, the world was shut down, along with the in-person events that were cancelled. I had to rely on whom I had already met and meeting new people online or at virtual events. I first went to my email list to check in with everyone and see how people were getting along. It wasn't so much about business at first:

it was about keeping connected to the relationships that I had already made.

The majority of my clients either met me in person at an event that I was attending or have seen me online at an event that I participated in or where I went live on a social media platform, and the video was shared and they were introduced to me. Most of my clients are not from my family, circle of friends, or those who knew me before I published my first book. Prior to writing and publishing my first book, I was a singer who sang gospel music in churches around the country and overseas.

Online Newsletter

One thing that I did back in the day that helped me greatly in the digital, online, and virtual space even today, is that I had a yellow notepad with me everywhere I went. If people wanted to stay connected with me, they would put their name, email, mail, and phone number on the pad of paper.

I would manually transfer their information to my email document and then copy and paste the email addresses into my business email and send each month.

The monthly newsletter that I sent out contained an inspirational message, what new music, and eventually books, would be coming out, and always where I would be singing next. I built that list to more than 900 people. I was so proud. The interesting thing is that when I started publishing books, I let people on my list know and if they wanted to unsubscribe they could. No one left the list.

Today, most business coaches suggest that you start from scratch and eliminate the people who were on your list before and interested in another business or venture that you had. I probably should have done that, but I didn't. Not me. There may have been a few people who unsubscribed themselves, but the majority stayed. The people on my list from music, travel, and concerts became my publishing publicity team, referral members, beta readers, helped me to pick covers, and always my cheerleaders. I still sang and published books for at least the first ten years of publishing, but now, publishing has taken over.

My point is, don't eliminate people, systems, and communication just because a coach or other mentor says so. These same people will unsubscribe themselves if they are no longer interested but unbeknownst to me, people don't have to buy from you to be on your list; they just may want to subscribe, share, and send your information to others.

The newsletter, whether weekly or monthly, became my first virtual communication space. I still do it today and utilize multiple email marketing systems to communicate with current and potential clients.

Email Marketing

In the virtual space, the email list and the weekly or monthly newsletter became even more important. I had built an email list, I had an email marketing system, and because of the pandemic, I relied on this list more than ever. Why? When the country shut down, I couldn't travel. I also didn't know when I would be travelling in the future. At the writing of this book, it has almost been a solid year since I have been at a major live, in-person event.

I followed some business coaches' advice and made sure that I connected to, checked up on, and encouraged people on my list. No selling at first, just "checking in on everybody." I then realized that people were home, and some people on my list had shared that they actually wanted to write, so I encouraged it. The campaign became "since we're home, let's write." I offered a writing class and had people sign up for every quarter in 2020. I had people reach back out who had delayed publishing their books and were ready to go. I still had people to refer me to others, and between the email list and social media, I didn't miss a beat. I realized that keeping in communication with people with a list on a consistent basis made it much easier to reach out in a crisis. I encourage people to start building an email list or list of people whom you know who are interested in your book or your business because I have found that it is much easier when things happen to have a group of people already connected to you than trying to reach out to people and striving to build relationships with a cold call or starting from scratch. Remember, people do business with people whom they know, like, and trust. That business relationship has to be built and may take longer than just one month, six months, or even a year, so if you don't nourish clients or potential leads for clients in a crisis, you may be hungry for a while.

Moving forward, I wanted to streamline how people connect to me because I wanted it to schedule discovery calls, get on my list, and be able to reconnect with in the future in one event. So no matter if you were an existing client, referred to me by an existing client, found me on social media, found my website, or saw another advertisement, there would be one way to connect via www.talkwithroyston.com.

I knew that working from home or in the virtual space, I not only needed to be streamlined but automated. I subscribed to another scheduling system that would link to my Google Calendar, and then it would automatically ask people if they wanted to be on my email list, and, if yes, it would add them automatically. This worked! I can't stress enough just how list-building has been key to growing and expanding my business, as well as adding value to potential clients prior to making a decision to publish or not.

Blogging/Articles

I stated earlier that I love to travel. I still do. Because of the digital, online, and virtual space, I needed to find more ways to not only promote but to also get me and my message of my business in front of new people. Writing articles for magazines and/or blogging to my own audience has been the answer. It depends on the specifications of the editor of the magazine, but you can use the same blog with a few tweaks for an article in a magazine. This is called repurposing content and something that I've become great at doing. I revive, refine, and refurbish old articles with new content all of the time. The ability to write hasn't changed, but the publishing and distribution methods change all of the time. So depending on the publication, I can tweak it to a specific audience all of the time.

Now, I didn't start writing articles during the pandemic and recent crisis, but it became more important during that time. It helped me to not only be in front of new audiences but also when a new magazine issue was promoted and I was in it, I could be seen on another social media and email list from my own. As a writer, it is a win/win situation. There are digital and online magazines everywhere. Do a search. Ask the editor if they are looking for articles. If they pay, great, but if they don't, use it for promotion, advertisement, and marketing. I have been able to participate in so many opportunities just writing articles.

If you need help, let's have a conversation at www.talkwithroyston.com or purchase the Write an Article Course at www.juliaroystonstore.com. Let's go!

Virtual Networking

Even though I am an extrovert by nature, going to network events was different for me in a room filled with new people and/or people I didn't know. The awkwardness of first, making eye contact and then, introducing myself. Yuck! But imagine doing that in the virtual space where people can quickly and more easily ignore you. On social media, they can block, unfollow you, and even report you. So how do you network in the virtual space?

Join a Group. Groups on Facebook and LinkedIn are the best ways to network in the virtual space.

Rule #1: Know the rules of the group. Each group has certain rules. Know the rules or get removed very quickly. There may be certain days to post. There may be certain times that they want comments. There may be the ability to promote, and then certain groups may not allow promotion of any kind. Know the rules.

Rule #2: Watch, learn, and listen. It only takes a few visits until you have learned the rules, figured out who the administrators are, and what the major issues are of the people in the group. Before you offer advice, give opinions, or offer anything, watch, learn, and listen. Of course, the listening can be either via video or just listen to the words that people are writing.

Rule #3: If you have a question or a request of the group and there is not a rule that doesn't allow it, ask the administrator first. Show respect and inbox the administrator first. It shows good character and it will pay off in the future.

Rule #4: If you have an opinion and there is opportunity, give it. Make sure that the opinion is on the subject and you are not trying to market and promote on the sly. Don't do it because they will kick you out.

Rule #5: Don't start a group from inside of someone else's group that they built. It is not good business. Enough said on that. Don't do it.

Reflection

_____

_____

_____

_____

_____

_____

_____

_____

_____

_____

_____

_____

_____

_____

_____

_____

_____

_____

_____

_____

_____

Virtual Audiences

Around the World in 10 Seconds or Less

In the summer of 2020, I was in-boxed, or sent a direct message, from a young man from India on Facebook. He was inviting me to an International Poetry Networking Event on a weekly basis. I couldn't be there every week, but the weeks that I could, I joined in at 9:00 a.m. EST. When I showed up the first Saturday, there were literally poets from around the world. From the East to the West. I couldn't figure out all of their time zones, but some of them had to have been logging on in the middle of the night, the late evening, or even the next day. It was incredible. Now, how did he find me? Online. Would that have happened in the in-person or non-COVID-19 space, maybe or maybe not, but I was able to network and meet people from around the world from the comfort of my own home without buying a plane ticket or hotel or suffering through the security check points in any airport in the world. It was amazing. I was humbled and honored. I am now going to publish one of my poems in an anthology with these global poets. I am super excited. The opportunity came to me and I'm thrilled.

My advice to you is this:

- Be open.
- Listen with new eyes and ears.
- Learn from the other people.
- Put your culture, backgrounds, and beliefs aside to see from someone else's viewpoint, culture, background, and belief system. It could change your life, expand your horizons and cause you to see the entire world differently.

For me, the digital, online, and virtual space has changed my business, my exposure, my audience, and my life.

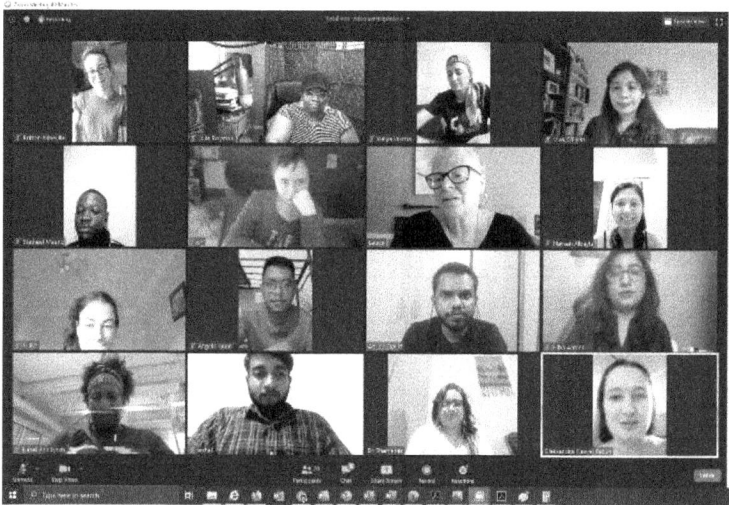

Reflection

_____

_____

_____

_____

_____

_____

_____

_____

_____

_____

_____

_____

_____

_____

_____

_____

_____

_____

_____

_____

_____

_____

_____

Virtual Events

At the end of 2019, I had already booked hotel reservations, bought plane tickets, paid for vendor spaces at large conference events, been invited to speak and conduct writing workshops at major business, church, and women's events in 2020. I was super excited because I have a client in London and we had been planning a women's conference for two years with the deposits on hosting hotel, room rates and the menu settled, as well as more than 60 people interested in going to London, England, with a side trip to Paris, France. I had big plans for 2020.

I had my travelling wardrobe bought, including new luggage, table coverings, promotional materials, website revised, inventory ordered and was ready to be on the move and in the air or on the ground in 2020.

Fast forward to March 13, 2020, when everything was shut down. The hotel reservations were cancelled, I still have coupons for plane tickets for 2021. Some events refunded my money and others kept the money because of a no-refund policy.

As I said earlier, I had seven clients immediately cancel their contracts. Thankfully, I had other authors who were so far into the publishing process that they said they had the money to finish and they did. I called some of my friends and family, and we strategized on what I should do. I prayed. I admit, I was scared of what would happen next. Would my business survive? Would anyone want to write a book or have the money to publish a book with me again? I cut my teeth and built my business on in-person events. People met me, liked me, and would do business with me, but did I have that same impact virtually as I did in-person? That was the question.

I have been teased by my fellow business friends that I'm the "Queen of Events" especially those events where I had to travel. I love to travel. Visit new places. See new things. Eat new foods. The whole experience is great to me. When my friend Vanessa Collins called me, she would always ask, "You home or where in the world are you?" I would laugh back then, but in March of 2020, I wasn't laughing. It was no longer funny. I had literally travelled from one end of the U. S. to the other for an opportunity to share my business and books with others. What happens now?

If I was panicking, I knew that my clients were panicking even more. So I had to create an opportunity for them to be comfortable going live about their books. They had invested in me, and I felt a sense of urgency to invest in them. It wasn't as if I had never gone live on social media or had never done an online or virtual summit or had never interviewed someone in the virtual space, that wasn't the case. It had been a balance. I did it when there wasn't another in-person event that I was travelling to, or it was winter time and I wasn't going to travel in the snow. Come spring and summer, I was on the road every single weekend.

What did I do first?

Because I'm in the publishing business, I hosted monthly book fairs for my authors. I asked them in my private group who was available, here is the log in, and then we went live. For National Library Week, I had a week-long book fair and spotlighted different genres of books each night. I had special events each month in addition to the book fairs.

I did "lunch and learn" events in the middle of the day to talk about the importance of writing and to ask people who were interested in writing their story to register for my 10-week courses. I filled the courses with students every quarter of 2020. People were home, why not write? Let's go!

I gathered like-minded business owners, authors, and other industry leaders in a Zoom room and went live on social media. Just like I travelled around the country in the physical space with tenacity, I did the same in the virtual space.

Anybody else's platform on social media or a virtual event that I was invited to, I showed up. I was the facilitator for other people's events online. I joined social media groups. I started a podcast. I went live on social media on a weekly basis or any other time, to stay in front of my friends, family, and potential clients. I turned my office into a dressing room because I would have multiple events on the same day, change clothes, jewelry, and lip gloss. I was committed to surviving this pandemic with my business intact. In the middle of it all, I taught classes, buried family members, saw friends suffering, as well as tried to comfort perfect strangers to keep going. I also had two surgeries, the flu, double pneumonia, four hospital stays and still published 50 books at the end of 2020. How? Only God knows. I willed myself to work, complete the books and fulfill my client contracts. The ultimate was being in the hospital with COVID with my laptop on my bed and oxygen in my nose, finishing up a client's book who had paid in full. She didn't pressure me, but prayed for me and said to take my time. I didn't listen. I was committed to it. Pushed publish to Amazon while getting a treatment. Yes, I'm a workaholic. I confess it, but looking back it was a catalyst to keep me going and ultimately stay alive.

People with an entrepreneurial spirit, find a way. It's not just about what you say, it's about what you do that counts even in the tough, rough, and terrible times.

As you can tell, this was a difficult time for me, but you don't know what you can do until there is trouble or a struggle.

JULIA ROYSTON ENTERPRISES PRESENTS

EDUCATIONAL SUMMIT 2020
AUGUST 10-14
7:00 P.M. EST - FACEBOOK LIVE

Host, Julia A. Royston

Dr. Wycondia West
August 10, 2020
Parent Summit

Dr. Shaniece Williams
August 11, 2020
Administrators

Dr. Edward Williams, Jr.
August 11, 2020
Administrators

Dr. Vernett Nettles
August 11, 2020
Administrators

Sakia Dixon
August 12, 2020
Social Services

LaTascha Craig
August 12, 2020
Family Resource Coordinator

Dr. Albert Brinkley
August 13, 2020 -
Educating African
American Males

Dr. Robert Gunn
August 13, 2020 -
Educating African
American Males

Desmond Williams
August 13, 2020 -
Educating African
American Males

Dr. Tamara Bowles
August 14, 2020 -
The Future of Education

Dr. Sharon Hunter
August 14, 2020 -
The Future of Education

Dr. Yvette Thomasson
August 14, 2020 -
The Future of Education

TELLING OUR STORIES
MEMORIES OF LOVE

Monday, July 13
7:00 p.m. EST

Julia A. Royston,
Host

Betty
Winston Baye'

Karen
Dickson-Brown

Cassandra
Lanier

Valerie
Johnson-Reed

Casilya
Smith

Elgina
Bullock-Smith

www.bkroystonpublishing.com

f LIVE

You Tube

PREPARING FOR
BODY, MIND AND BUSINESS
Virtual Summit
2021

FELITA POOLE

VANESSA COLLINS

STACI STILLS

LASHAUNDA HOFFMAN

Julia A. Royston

RITA OLIVER

SAKINAH BUNCH

DEC 14, 2020
7:00 P.M. EST

Facebook LIVE, Youtube LIVE and Periscope
Register at http://www.juliaroyston.net/bodymindbusiness

Keys to Virtual Events:

Be strategic about the topic. Are people wanting to talk about this topic? The Internet is crowded with all types of events. Make the topic relevant.

Be careful of when you schedule the event. Days and times do matter. Days and times that are convenient for you may not be convenient for others. When I went live on social media early on in the pandemic, everything was closed and people were home, but as the country began to reopen, I had to reconsider the day and time that I went live.

Be careful of whom you select to be a guest. Make sure that they are also active on social media. The ability to get people to tune in is predicated on people knowing about the event on other platforms and not just the hosts.

Be careful to not make every event a one-sided endeavor. Be sure to share other people's events just like you want people to share your event. Be sure to promote your guest's websites, contact information, etc.. It can't just be about your business, book, or interest but everyone who is participating. If you are not comfortable with this, don't invite others: just go live on social media alone.

Graphics. Provide a quality flyer with the pertinent information and outlets that you will be streaming.

Try out new platforms. We like our comfort zones even in the virtual space, but I am open and learning new platforms all of the time. Some platforms will eliminate themselves because of budget, but others will have free or trial memberships that will allow you to explore, expand, and extend your reach of making even more connections to people around the world.

Practice, Practice, and Practice some more. New software, features, and anything new that I want to teach, I invite a small group of people whose opinion I trust and know that they have been loyal clients and invite them to try it out. I ask for their honest opinion and take their suggestions into consideration. Prime example, I use Zoom, but Zoom offered Zoom rooms to be able to separate discussions. Prior to offering this or orchestrating this as a host of a virtual conference, I tried it out first. I wanted to see how it would work. Practice does make, not perfect but more comfortable.

Promote, Promote, Promote. Enough said, but it has to be said. As soon as you have a flyer, share it. In the virtual space, one to two weeks is fine if it is a free event. That works. If it is a paid event but under $50, give yourself 30–60 days. If it is a paid event more than $50, you are going to need more time to promote depending on the topic. Timing is everything. Don't wait too long because with technology and the swift changes in our world, there could be a more pressing topic to talk about and your topic will be old news. Move quickly and strategically. Don't wait. Move.

GO Live no matter who shows. Why? The replay. The views later can be incredible to moving your message to new audiences.

The virtual space has afforded me the opportunity to do business with, meet, and participate in events around the world that would have taken thousands of dollars and maybe years to create that same opportunity. With the virtual space, I can literally be anywhere in the world any time of day or night.

I believe that the virtual opportunities are here to stay and there will be more of a balance once the in-person events, businesses, and opportunities are safe to return. Will some people never go back into the in-person quickly and easily unless absolutely necessary? I am certain of it.

For more information or to have a coaching or mentoring session concerning your virtual endeavors, visit www.talkwithroyston.com.

Below is the first small ad regarding the first Virtual Book Fair. I only let 11 days go by before I had my first book fair.

Below is my email header for my April newsletter, "We're Home, Let's Write!"

www.talkwithroyston.com

We couldn't go to London. So we went virtual instead, with the speakers who were scheduled to speak at the London, England, conference.

BK ROYSTON PUBLISHING AND HEART OF REFUGE PRESENTS

EMPOWERED TO SOAR

Virtual Summit
Saturday
April 25th, 2020

ISAIAH 40:31 - BUT THEY THAT WAIT UPON THE LORD SHALL RENEW THEIR STRENGTH; THEY SHALL MOUNT UP WITH WINGS AS EAGLES; THEY SHALL RUN, AND NOT BE WEARY; AND THEY SHALL WALK, AND NOT FAINT.

12 Noon EST
5:00 p.m. London Time

SPEAKERS:

JENNY ALLEN   MONICA AMBERSLEY   VANESSA COLLINS   JULIA ROYSTON

Free but Must Register for Gifts, Give-aways and Replay at:
http://londonvirtualsummit.gr8.com

Conclusion

Create. Change. Clients. Repeat.

I am a creative person, so I stay in create mode. The weakness of constant creation mode is that you don't analyze as thoroughly as you should or could because you are busy creating the next thing. This is good and bad. I may resemble the first statement, and others may resemble this next statement. You can't be analyzing results so long that you forget to create something new or in a different format that your customer now wants versus what you offered them before.

Even if I have a customer who wants what I have to offer, I also package it and will launch it later with a broader marketing and promotional effort for someone else, somewhere else, to see it and buy it. I don't think any of my products and services are wasted; I just need to make sure that it gets in front of the customer who wants, needs, and is ready to pay for it.

On the other hand, I do have to take more time than twice a year or when something goes wrong and doesn't turn out right, to look at my production, ventures, and book launches to tweak some unsuccessful things, create something new, and/or celebrate the things that actually did go well. It's a balance.

The pandemic taught me to determine quickly what's important, decide a direction to head down, and, most importantly, implement. It wouldn't have made as much difference in my business and the books of my clients if I didn't reach out, decide the date, design the graphics, and push the GO Live button. Call it what you want, "shifting," "pivoting," "changing," or "making adjustments," just do it to keep moving forward in a positive direction. The sacrifice paid off not only for me, but also for others.

In 2021, I will probably not host as many book fairs as I will host special topical events. Why? Last year's authors are now conducting their own events, going live on social media on their own, and have seen a "blueprint" of sorts and are launching out on their own. This is a sign of growth, and I'm so proud and happy I could bust.

Now, I can go from panic mode to planning mode. I can go from being scared to being strategic. My business, just like the world, is not fully out of the woods yet, but at the writing of this book and for those of us who survived 2020, the future is as bright as it will ever be. More importantly, we have a sense of the highest priorities and that is faith, family, friends, and staying focused.

Reflection

_____
_____
_____
_____
_____
_____
_____
_____
_____
_____
_____
_____
_____
_____
_____
_____
_____
_____
_____
_____
_____
_____

Business Lessons Learned

- ❖ Always do what you love, and when you love what you're doing, you will find a way to do it.

- ❖ Don't let your dream goal vision impoverish you; always fund your dream with a full-time job, a part-time job, or another stream of income.

- ❖ If the business is profitable but you don't have the heart for it, it will still impoverish you. You will find yourself not working as hard for your business as you do for a job because your heart is not in it.

- ❖ Your reputation, your name, your ability to produce is your brand and not necessarily the colors, the logo and the color scheme.

- ❖ Be willing to cut ties with things that are unsuccessful and people who are toxic.

- ❖ You can always make more money but rebuilding a reputation takes forever.

❖ Every client is not your client, but there is somebody out there who wants to work with you; do you want to work with them?

❖ Remember your family and friends may not always be your clients, but they may be your tribe, team or they may be your volunteer staff until you can pay for staff, but do not look for family and friends to be your only or first clients or to buy from you.

❖ No matter the relationship, find the balance; otherwise, find a new relationship.

❖ You are your brand and your brand is you. When doing business, be careful whom you link your brand to. The short-term money may be great, but you could spend a lifetime untangling, separating, and rebuilding your reputation as somebody that people want to do business with, because they know, like, and trust you.

Reflection

_____

_____

_____

_____

_____

_____

_____

_____

_____

_____

_____

_____

_____

_____

_____

_____

_____

_____

_____

_____

_____

_____

P. S. Year Three Lessons

I thought I was done with the book and ready to go to the editor, but some other thoughts came to mind.

Balance — I have to drive home the "balance" word one more time for myself, I believe. Find the balance between business, personal life, rest, taking breaks, meetings with clients and when to just be alone with your thoughts. For example, scheduled meetings with clients are necessary for income, but if they don't show after 15 minutes, I shut Zoom down and move on to my next task. I used to hold up myself and my day, begging people to meet. Things happen and emergencies come up, I get it, but I have to move forward. I wasn't okay in year one of business or year one of full-time entrepreneurship, but a business associate said,

"Girl, if you don't pace yourself, you'll burn out."
She was right.

Habits — "Going hard in the paint," has been my "go to" phrase for the first three years of full time entrepreneurship journey, but I am committed to changing parts of my life so that I don't have to go back to a regular job ever again. First, to save time and gas, I group my travel events if possible. I don't go out unless I can accomplish a multiplicity of errands and tasks. I'm on a better schedule, and I don't waste time and money.

Second, this may sound gross, but since I'm home and not around as many people, I don't use as much cologne or extra smell good stuff. Now I bathe every day and do put deodorant on, but not as much as I did when I worked outside of the home. I don't get manicures, pedicures, and massages every month as I did when I worked outside of the home because of the expense. I'll go a few extra weeks with my toes not looking so great because who is seeing them in winter but me, my husband, and my sneakers? I get my hair done, not every week but every other week. I have even purchased wigs to change up my hair style when the country was on lock-down and the hair salons were closed, but I could still look presentable online. I still want to look professional, but since I'm home and a full-time entrepreneur, I must watch my habits closely. More importantly, I am willing to change my habits to keep the lifestyle that I want to continue and grow. I believe in self-care and take care of me, but there is a limit to the amount of expense and time placed on self-care. It's a choice but I'm not willing to risk my business for time in the salon unless absolutely necessary. It's a choice and I choose me and my business and not my pampering over having funds that will fund my self-care and my future.

Spending — Since I work from home and even when I was working outside of the home and planning for my business, I curtailed a lot of wasteful spending. Why? I needed the funds to invest in my business. I love to shop but do I spend $100 on a new dress that isn't necessary or take that same $100 on advertisement to move my business further? It's a choice. I now eat more at home and eat my husband's wonderful cooking. Now I love to eat out, too, sometimes but don't want to spend the money for eating out when I can eat something just as wonderful out of my own refrigerator. I have plans for my money. It is year three of my strategic plan, so I have to be as strategic with my time as much as with my money.

My Team — I talked about having and hiring a virtual assistant in an earlier chapter. Make a note of this in your planner and when you have the budget for it, figure out the part of your business that can either be outsourced and do it. The biggest reason is not to appear like a big shot or go on an ego trip but so that you can work on your business rather than so much working in and at the work of your business.

Professional Development — Take time to learn something new or improve on what you already know. As a teacher, I was required to participate in and gain certificates for 24 hours of professional development a year. I don't get in as much time as I want, but I am a webinar attending, eBook downloading, and live stream watching Queen when I have time. Don't tell me that you are providing the replay, sign me up! You might be laughing, but I'm not. Sharpening your skills is so important. Let's go!

Faith Workouts — Entrepreneurship is hard enough with God but no matter where your level of faith is, put your faith in something that is eternal and won't fail. That is all.

Reflection

_____

_____

_____

_____

_____

_____

_____

_____

_____

_____

_____

_____

_____

_____

_____

_____

_____

_____

_____

_____

_____

_____

About the Author

Julia Royston spends her days doing what she loves, writing, publishing, speaking, and coaching others to tell, introduce, and create ways to deliver their stories and messages to the world. That is her "Why." BK Royston Publishing LLC, Julia Royston.net, Royal Media and Publishing, Royston Book Fairs, and BK Royston Foundation are the conduits that she and her husband, Brian Royston, use to spread the love of reading, writing, and books, as well as build businesses around the world. Julia has written 55 books, recorded three music CDs and coached 150+ to write and publish books, as well as establish their own businesses. Julia is the host of "Live Your Best Life" heard every Sunday morning at 10:00 a.m. on www.envision-radio.com. In 2020,

BK Royston Foundation was established to address literacy, diversity, and inclusion issues in the global society.

For more information, visit www.bkroystonfoundation.org.

Follow her on social media

Facebook, LinkedIn and Instagram @juliaaroyston

TikTok — @juliaroyston

Twitter — @juliaakroyston

Visit the websites
www.bkroystonpublishing.com, royalmediaandpublishing.com, bkroystonfoundation.org or www.juliaroyston.net for more information and upcoming events.

www.ingramcontent.com/pod-product-compliance
Lightning Source LLC
Chambersburg PA
CBHW071851200326
41519CB00016B/4331